Anton Furst
Nick Phillips
John Wolff

Bergström+Boyle Books Limited, London

"Science" wrote Lethaby fifty years ago "is what you know. Art is what you do". Science can test, codify, and inform. It cannot choose. Hence the contribution of the artist is crucial and the partnership of knowing and doing truly indivisible.

This is why we at the Royal Academy welcome so warmly in this exhibition the evidence of such partnership. It is the cooperative venture of a physicist (Nick Phillips), a specialist in the optical effects of laser beam technology (John Wolff) and a film art director (Anton Furst) – working together under the name of "Holoco".

The exhibition has been devised to demonstrate, virtually for the first time, the aesthetic opportunities of the image-forming techniques made possible by the development of Holography – invented in 1947 by Professor Dennis Gabor at the British Laboratories at Rugby – and of the laser beam.

These opportunities, hitherto confined to the laboratory and a handful of public experiments, are beginning to make possible a new visual vocabulary at the service of science and art alike. The Royal Academy, which since its foundation has existed to serve and nourish the arts, is glad to act as host to so unique and potentially exciting an enterprise.

Sir Hugh Casson
President of the Royal Academy

LIGHT AND LASERS

Galileo (1562–1642), regarded as the father of modern physics, first attempted to measure the velocity of light but concluded that its speed was infinite.

Light is a form of radiation. Because it travels so fast, the speed of light does appear infinite to the naked eye, but light is in fact made up of waves and is continually on the move at a constant speed.

The Dutch astonomer Ole Roemer, by studying the eclipse of the moons of Jupiter, later deduced that light did have a measurable velocity.

It has since been estimated that light travels at a speed of 186,000 miles per second, or, as the metric system is more commonly used nowadays, a speed of roughly 299,780 kilometres per second. Whatever the system of measuring employed, it is still a speed virtually beyond human comprehension.

The velocity of light is always and everywhere the same and for this reason is regarded as one of nature's constants, represented in science by a little c. It is a common symbol in physics. Those who have studied the subject will know that it occurs in the historic answer of Albert Einstein (1879-1955) to the question: "What relationship, if any, exists between mass (m) of matter and the energy (E) associated with that matter?"

He concluded that the amount of energy is equal to the amount of mass times the square of the velocity of light. For the equation-minded such an answer is written simply as $E = mc^2$.

It was the English physician Dr. Thomas Young who discovered at the turn of the last century that light travels in a wave motion, in similar fashion to the waves of the sea, the crests and troughs rising and falling in a direction at right angles to the direction of travel.

A century earlier the most famous of English scientists Sir Isaac Newton (1642–1727), born the same year as Galileo's death, succeeded in splitting white light into the various colours of the spectrum. Light is usually thought of as being white, but white light is nothing more than an amalgam of the colours of the spectrum.

Newton was a resourceful experimenter with a deep interest in light and optics, and by using a glass wedge or prism was able to separate beams of pure white sunlight into the differing colours of the rainbow.

It was Newton's conclusion that there was no such thing as a single light, but that there were many, identical in velocity, but differing in other respects. It is these differences that our eyes perceive as colours.

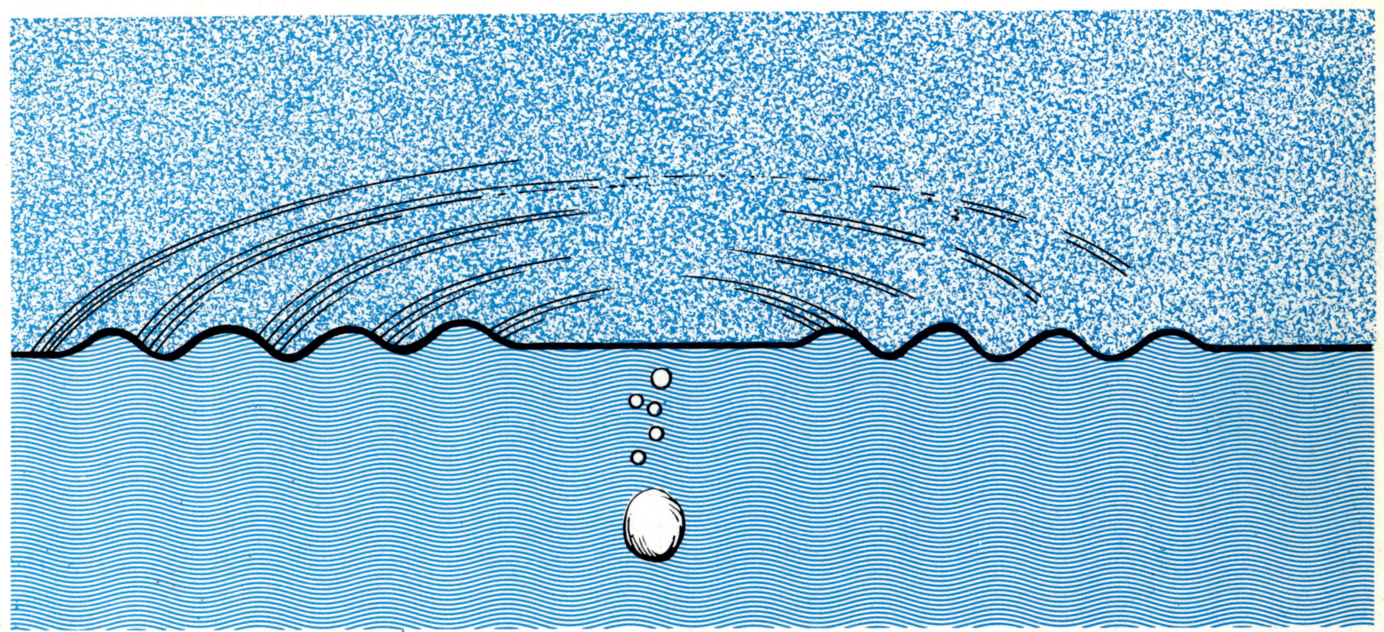

A stone thrown into a pond produces waves that behave in much the same way as light and sound waves.

A wavelength is the distance from one crest to the next and every colour has its own individual wavelength and also its own frequency. As in the case of all forms of radiation, frequency times wavelength equals velocity. It therefore follows that frequency must equal velocity divided by wavelength or that wavelength equals velocity divided by frequency. The shorter the wavelength, the higher the frequency and vice versa.

Normal everyday light, because it is composed of many differing wavelengths and frequencies, is regarded by scientists as being "incoherent" light, that is light with no regular pattern. Before the discovery of the laser, science had not found a source of pure light.

A coherent beam of light is one composed solely of waves of identical length which give the beam a regular frequency. As yet there is actually no such thing as a pure coherent light source, but the laser comes close enough.

The word laser means "Light Amplification by Stimulated Emission of Radiation" but in slightly more simple terms, a laser is a light amplifier with very special characteristics.

It was Niels Bohr's atomic theory that helped bring about the development of the laser. While investigating the relationship between the electron and the nucleus of the hydrogen atom, the Danish physicist discovered in 1913 that by exposure to outside sources of energy electrons could be raised to a higher energy level.

He discovered, however, that not just any energy source would suffice. It had to be equal to the energy difference between the "ground state" of the electron, that is when it was circling closest around the nucleus, and its "excited state", when the electron carved a circle of much greater circumference.

For the laser to work it depends on the very special emission characteristics of certain atoms whose electrons have been raised to an excited state. Once in an excited state the electrons give off energy as they fall back to their original lower energy level. This energy is emitted as visible light.

The most common laser today is the helium-neon laser which consists of a laser tube containing roughly 10 per cent helium gas and 90 per cent neon gas, the latter being the active agent in the lasing process. At one end of the tube there is a fully glazed mirror, at the other a partially glazed one.

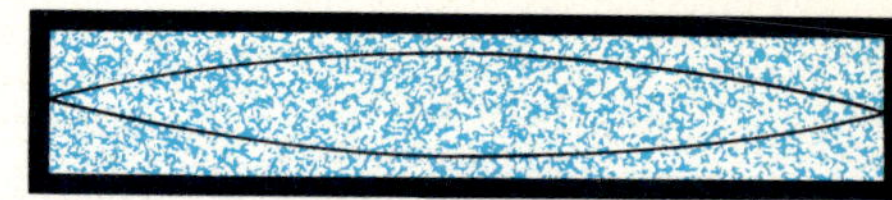

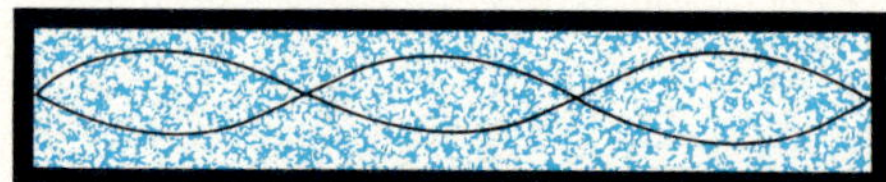

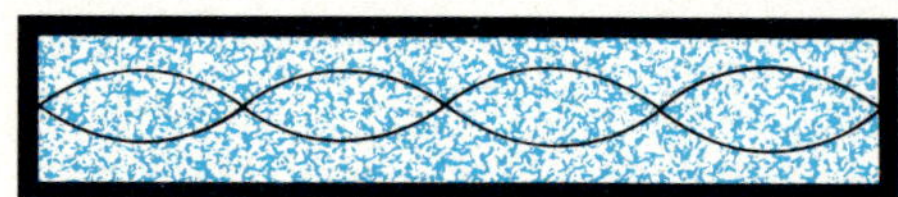

Colours have their own wave lengths and frequencies as do sound waves. Here different sound wave patterns are shown reflecting off the ends of a resonator similar to an organ pipe.

Scientists have discovered it is much easier to raise helium electrons to an excited state, and it is for this reason that the gas is added to the neon as a means of helping to raise the neon electrons to a similar state.

The electrons in the tube are fed with energy by a power supply. As they are fed the excited electrons start to outnumber those in a ground state, the result is an excess of energy in the excited state.

Ultimately as the excited electrons bounce up and down in between the mirror ends of the tube, they set up a regular optical wavelength pattern and frequency. There is, however, nowhere for the light produced to go except out through the partially reflecting mirror, and it is through this that the waves are emitted as a band of light with an almost regular wavelength. It is because of the regular, equal length of the waves that the laser is such a useful and powerful source of light.

A laser beam bouncing first off a building and then off a window. Some of the light goes through the window and illuminates the room inside.

It was in 1958 that American scientists Charles H. Townes and Arthur L. Schawlow published a paper heralding the coming of the laser entitled "Infra-red and Optical Masers." The two were brothers-in-law and had worked together at Columbia University, New York, where Schawlow was a postdoctoral fellow and research associate before he moved to the Bell Telephone Laboratories. Townes was the father of the maser, a microwave amplification device.

The paper proposed that atoms of a metallic vapour such as sodium or potassium could be pumped to excited states and then stimulated to emit coherent light radiation. It also laid down the basic design for a laser.

Schawlow suggested the laser action might be achieved by using a solid active material such as a ruby crystal.

The birthplace of the laser was the Hughes Aircraft Company Electronics Research Laboratories, which nestle in the hills overlooking the famous Malibu beach on America's Californian coast.

Laser equipment in action during a light show.

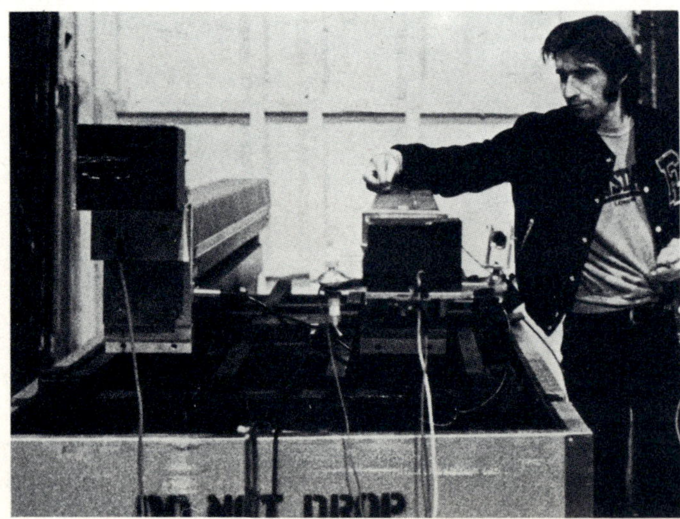

Laser equipment.

In July 1960 Theodore H. Maiman became the first man to achieve laser action. He did so by using a small cylinder of pink synthetic ruby about half a centimetre in diameter and a few centimetres long, pumped by powerful bursts of light from a flash tube, similar to those used by photographers in "strobe" lighting.

The duration of Maiman's red laser flash was extremely brief, about 3 thousandths of a second; but it was a very intense light, at its peak the flash attaining an estimated power level of 10,000 watts.

The invention of the Pulse Ruby Laser – a solid state laser with a synthetic ruby rod consisting of aluminium oxide mixed with a small amount of chromium – stimulated many scientists into entering the field of laser technology; and various solid state lasers were constructed using materials other than ruby.

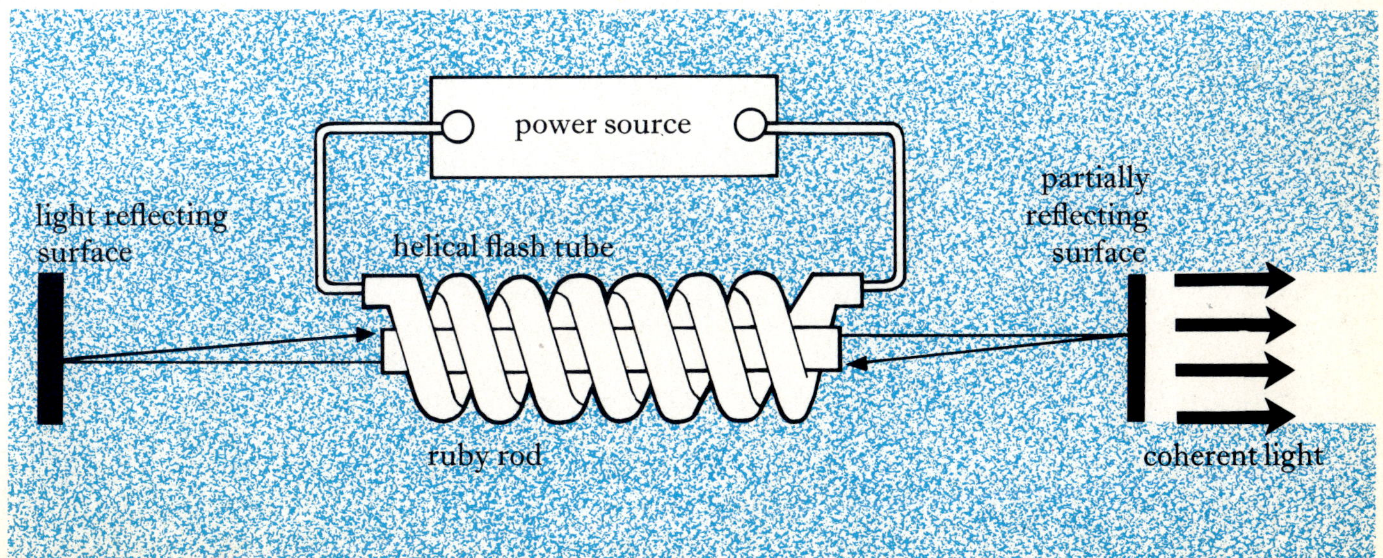

The essential components of Maiman's early pulse ruby laser.

But the real alternative to the Pulse Ruby Laser is the Continuous Wave Gas Laser. This as its name implies emits a continuous beam of light that can be turned on and off like a light bulb; whereas the ruby laser emits its energy in pulses.

Continuous wave gas lasers are not as powerful as the ruby version but are the ultimate in coherence and uniformity. The helium neon laser is the most commonly used of the gas versions if for no other reason than that it is the least expensive.

Light pattern produced by a Krypton Laser.

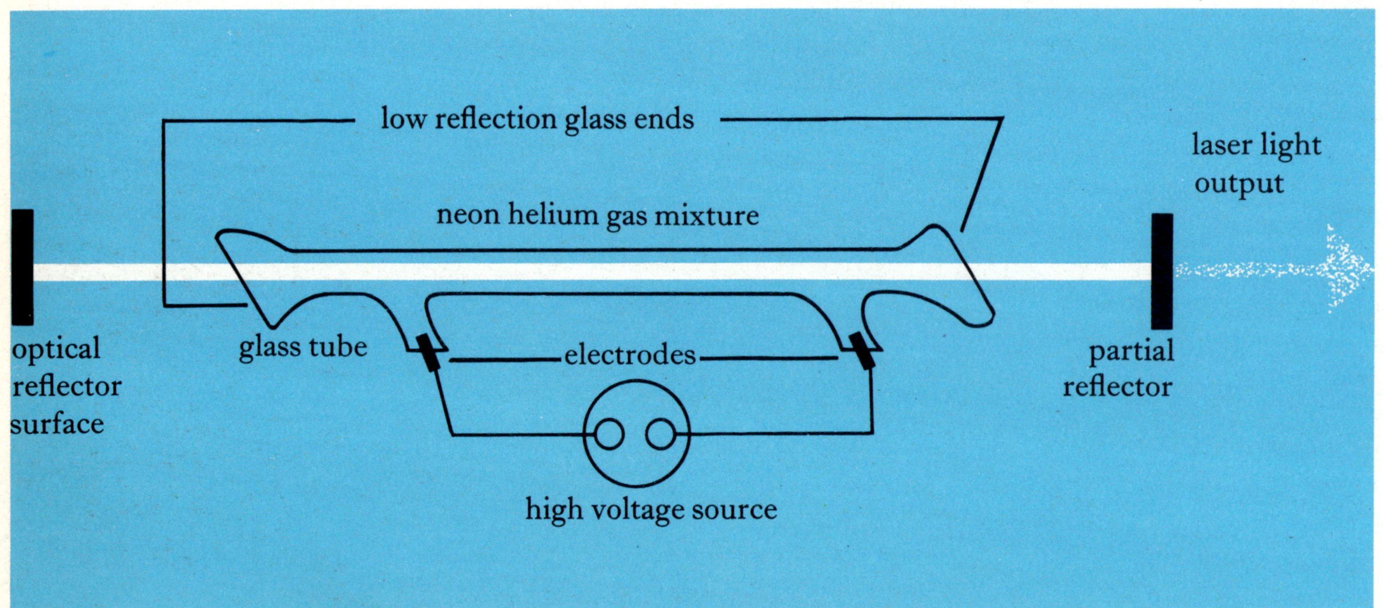

The basic features of a Continuous Wave Gas Laser. The two reflectors in the laser act as the end walls of the sound resonator.

A more powerful and more expensive Continuous Wave Gas Laser is the Krypton-Argon Ion version which emits beams of light in the blue, green and red regions of the spectrum.

Another is the Argon Ion version, capable of emitting light in various wavelengths of both blue and green light. This is also more powerful than the helium-neon version.

Without lasers the development of holography would have been impossible.

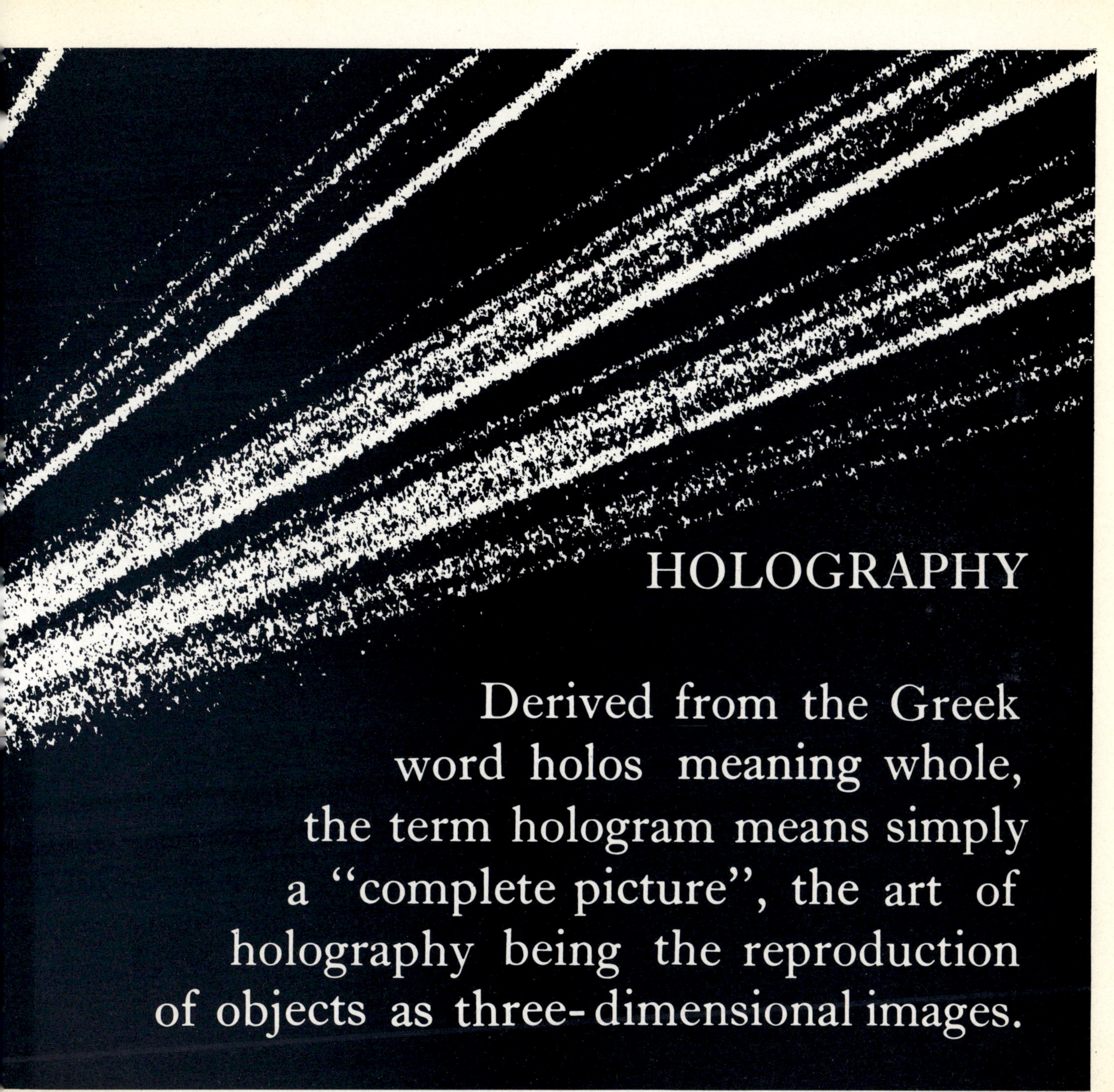
HOLOGRAPHY
Derived from the Greek word holos meaning whole, the term hologram means simply a "complete picture", the art of holography being the reproduction of objects as three-dimensional images.

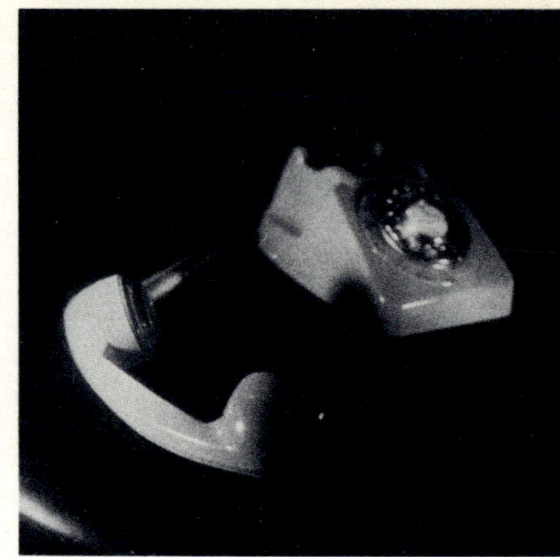

This series of photographs taken of a holographic image of a telephone shows how the relationship of the telephone to the handset alters from different angles.

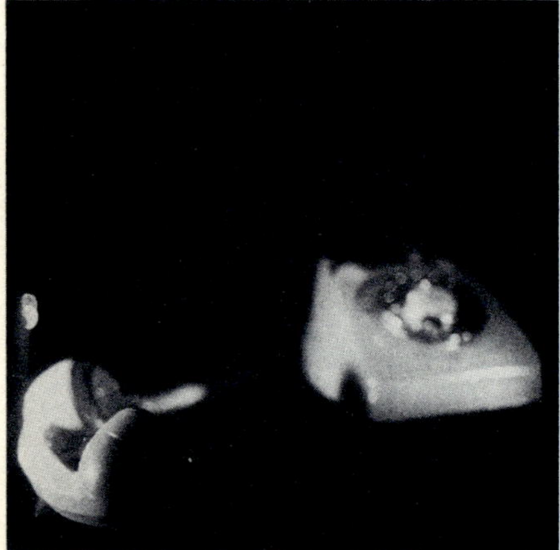

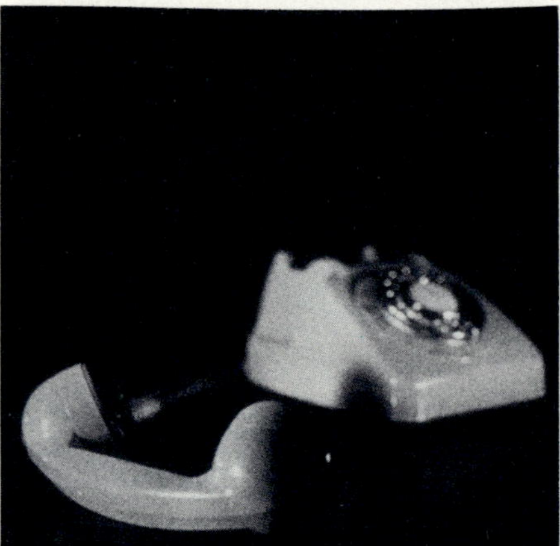

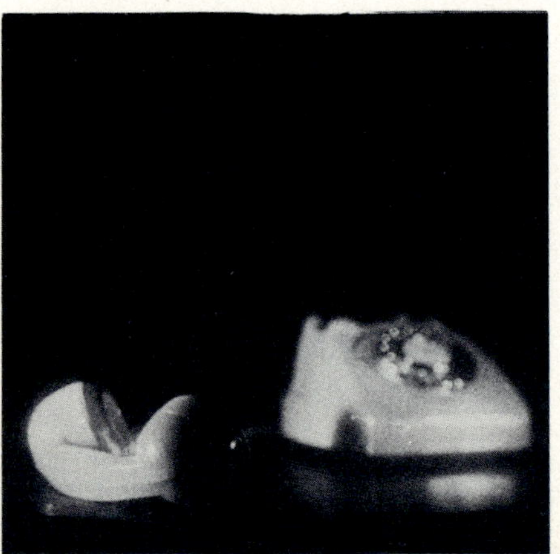

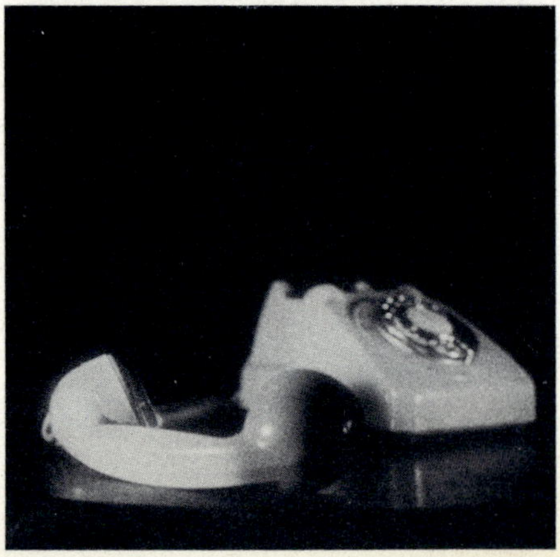

In many ways holography is a natural development of photography, although there are sufficient fundamental differences between the two to make comparison difficult. A photographic image is two-dimensional whereas in holography the image produced of the original subject becomes a three-dimensional form.

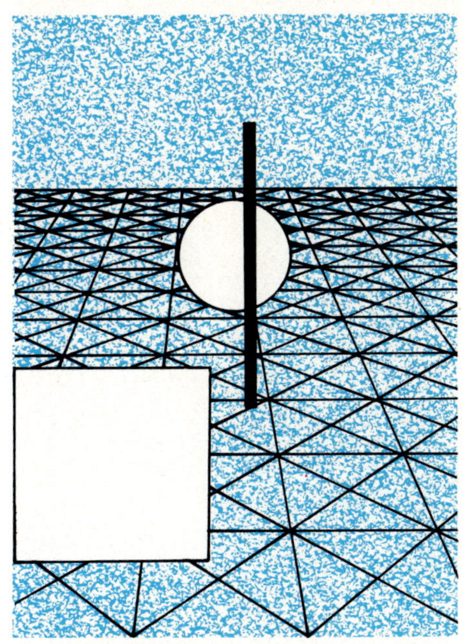

Despite the illusion of three dimensions through perspective, two-dimensional pictures like the one above will not change with different viewing positions. If it were a hologram it would behave like a real scene. By moving to the left you could see the circle unobstructed by the pole. From a lower position the square would obscure the circle. This is a simple demonstration of parallax.

A strong source of pure light, such as that generated by a laser, is essential for the making of holograms, whereas in photography normal daylight or flash bulbs are perfectly adequate for taking photographs.

Almost thirty years have elapsed since the Hungarian-born scientist Dennis Gabor first discovered the theory of holography, while working at the British Laboratories at Rugby. In 1948 he described how it was possible for an object to be reproduced in three-dimensional form apparently by the simple use of a light sensitive (photographic) plate. Twenty-three years later his discovery won him a Nobel Prize in Physics.

But Gabor was unable to take his invention very much further owing to the absence of both a sufficiently strong source of light and of a high quality emulsion suitable for his light sensitive plates. By using a mercury vapour lamp, about the strongest source of light available during the late forties and fifties, he was able to produce dim, blurry images that demonstrated his theory worked but aroused little interest in the invention. As in the pioneer days of photography almost a hundred years earlier, there was no shortage of sceptics eager to brand Gabor's holography an invention without a future.

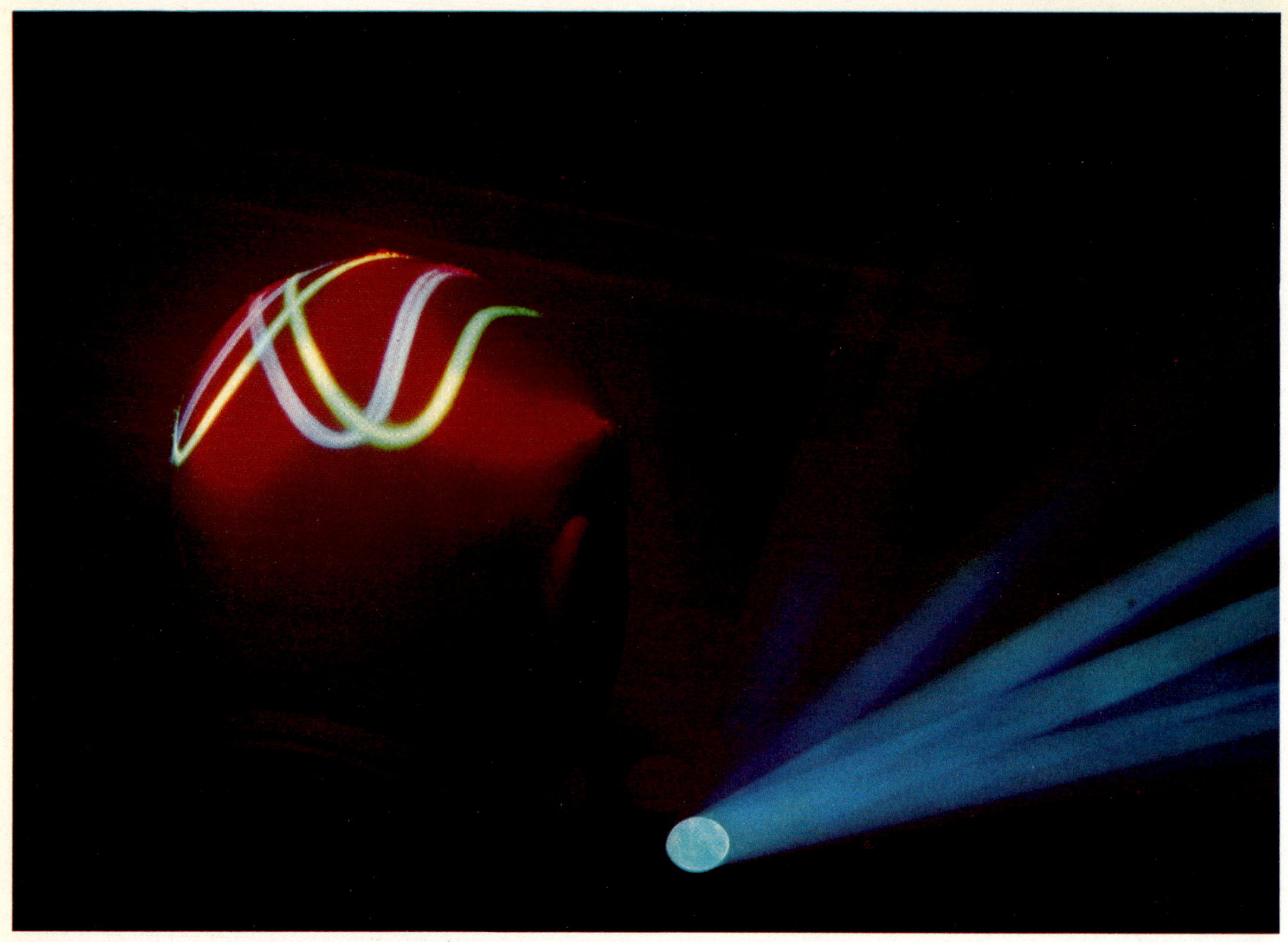

John Wolff: the brain behind the laser light show.

Interest was rekindled by the advent of the laser. At last a source of almost pure light had been discovered, so strong that it was capable of illuminating all objects sufficiently for sizeable holographic recordings to be made on light sensitive plates. Emmett Leith and Juris Upatnieks, two scientists working at the University of Michigan, USA, reproduced Gabor's earlier experiments using a laser and launched modern holography.

But while objects could now be recorded accurately on the light sensitive (holographic) plates, the image reproductions were still poor due to the absence of a suitable high quality emulsion. Since holography had appeared to have no future before the discovery of the laser, work on improving the quality of emulsions had ground to a halt.

Holographic plates are coated with a layer of gelatine and tiny crystals of silver bromide in which is dissolved the dye blanket that makes them sensitive to light. Finding the correct solution proved a painstaking operation, involving many hundreds of trial and error experiments, and it is only over the past year that a suitable emulsion has been finally perfected. The bulk of the film made for holography is currently being manufactured by Agfa-Gevaert of Belgium, who see the development of good emulsions suitable for work in this field as a long term project. Today the holographic images being produced in this country – and Britain is in the forefront in process technology – are as clear and at least as bright as the original subjects.

Professor Dennis Gabor, the inventor of holography. Photograph by Godfrey Argent.

Until recently those experimenting in holography possessed little confidence in the results of their work. Now the technique has been mastered to the point where results are virtually guaranteed, confidence is no longer in question and holography is for the first time being seen as a potential commercial proposition.

Although considerable specialist scientific knowledge is required to understand the technology involved in holography, it is not difficult to understand the processes involved in making a holographic image.

A photograph is made by recording on light sensitive film the differing intensities of light reflected by the object and focused on the film by means of a camera lens. In the case of holography, because it is a three-dimensional medium, the process is more complicated. The beam of light generated by the laser is first split in two. One part, referred to as the object beam, is directed by optical mirrors and spread by a lens onto the subject under study. As mentioned earlier, all objects reflect light and the light waves emitted by the illuminated object are reflected on to the photographic plate. The waves vary in intensity in accordance with the shape and surface nature of the subject.

Beam splitter.

Beam splitter.

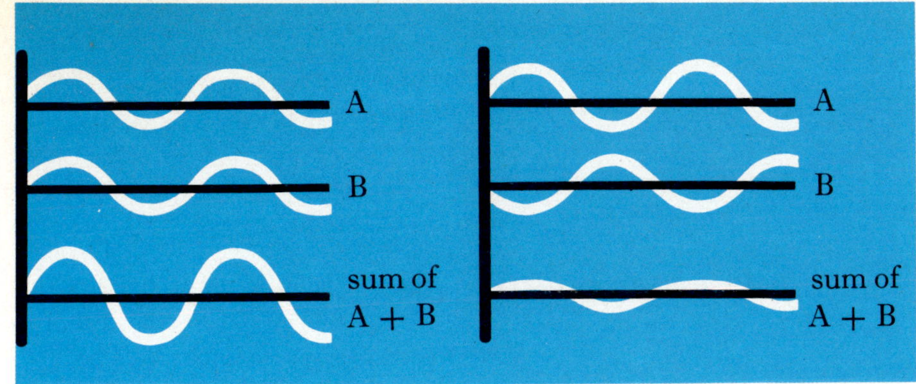

On the left two waves of the same wavelength add if their crests and troughs coincide. If the crest of one coincides with the trough of the other they substract (right). The diffraction grating of a hologram is built up by recording this sort of information.

The other part of the main beam, the reference beam, is directed by mirrors on to the photographic plate. This beam records the dimensions and depths of the object which make possible the three-dimensional effect of holographic images. Where the reflected waves from the illuminated object meet the waves of the reference beam, they overlap to set up an interference wavefront. It is this which is recorded on the plate.

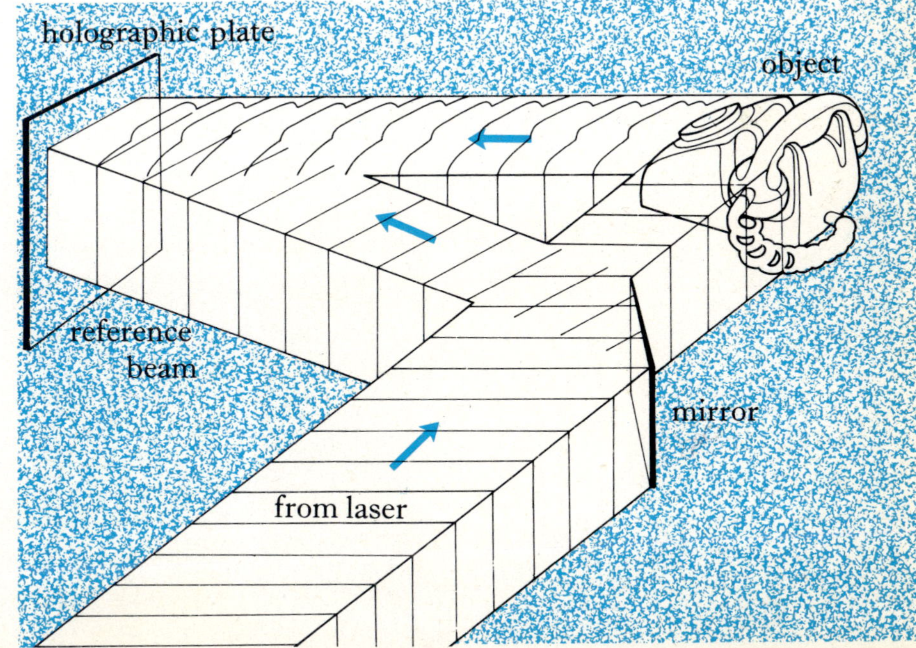

A projection of the process of making a hologram. The hologram plate records the complex pattern caused by the interference of the two beams. This can be called the diffraction grating.

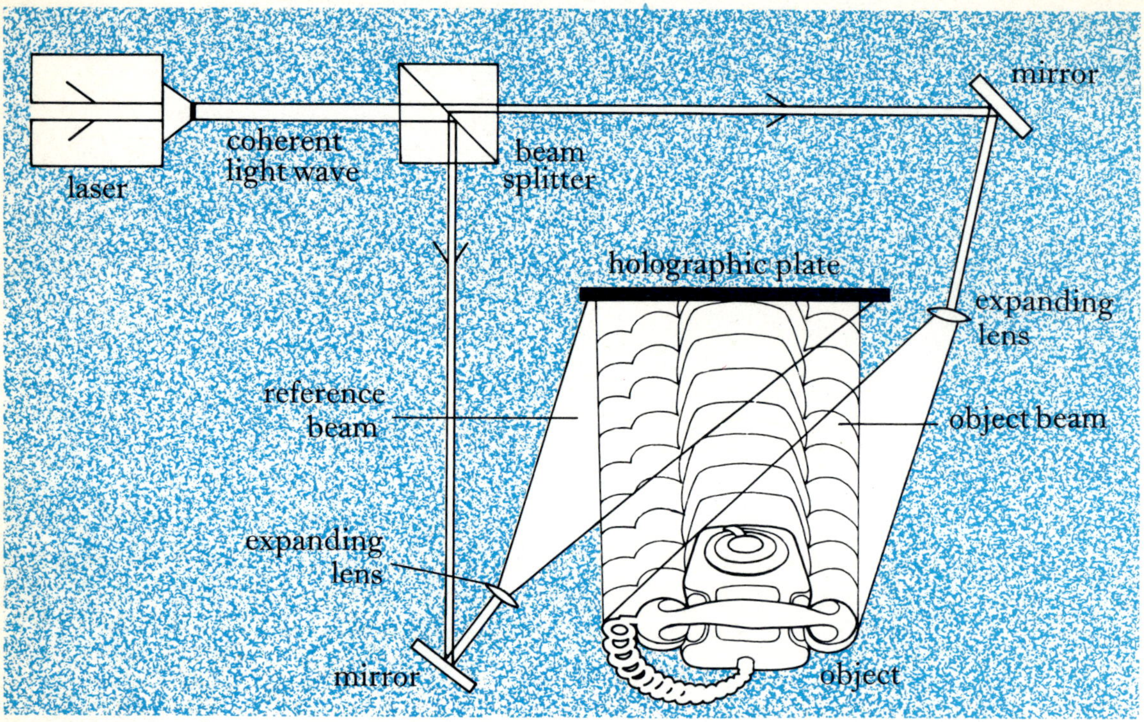

laser
coherent light wave
beam splitter
mirror
holographic plate
expanding lens
reference beam
object beam
expanding lens
mirror
object

Typical plan of a holographic recording showing the beam split to illuminate the object (object beam) and the hologram (reference beam).

Recording a hologram of a telephone in the laboratory at Loughborough University. The screens have a black surface which absorbs unwanted light.

The plate is made most commonly of glass since it is optically sound, an easy and stable material with which to work, and relatively inexpensive. There is, however, no reason why other materials such as film, acetate or Perspex should not be used so long as they are capable of being properly coated with the photographic emulsion. Sizeable glass plates are heavy and there are obvious advantages to be gained from using lighter and more easily transportable Perspex plates, or acetate film.

The slightest movement of the subject during the recording of a hologram does not blur the image but completely obliterates it, since it is the recording not of a focused image but of the interference of two wavefronts of light – the reference and object beams. A Continuous Wave Laser is used when taking recordings of static objects; but should the subject be moving or the process be carried out away from a laboratory, then a Pulse Ruby Laser has to be employed.

The time needed to expose a hologram – the correct term for the photographic plate – is dependent on a number of factors such as the power of the laser, the sensitivity of the emulsion and the amount of light reflected by the subject. The average exposure for a common hologram is anywhere between a second and a minute, although exposures of anything up to twenty minutes are not unusual.

The closer the subject to the plate, the nearer the holographic image will appear to the hologram when subsequently viewed.

After being exposed the plate is developed in the same way as normal photographic film and then a reverse process to fixing, called bleaching, is carried out, in which it is the silver that is dissolved away leaving the bromide crystals. The result is a plate with a complex diffraction grating. There is however no image to be seen on the hologram plate, which appears transparent until it is reconstructed.

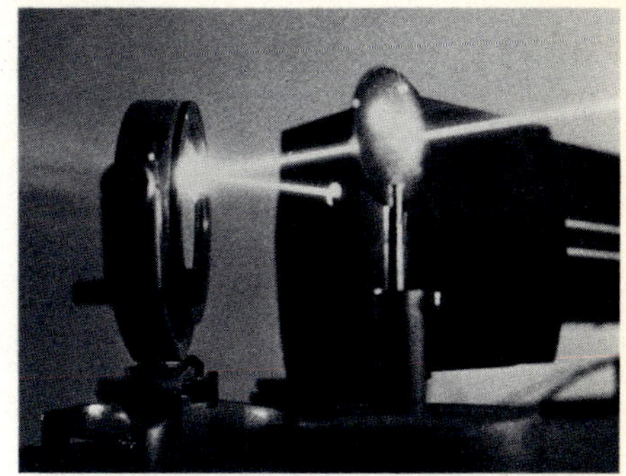

The laser beam is redirected by a steering mirror through a mask which stops any reflected light returning to the mirror.

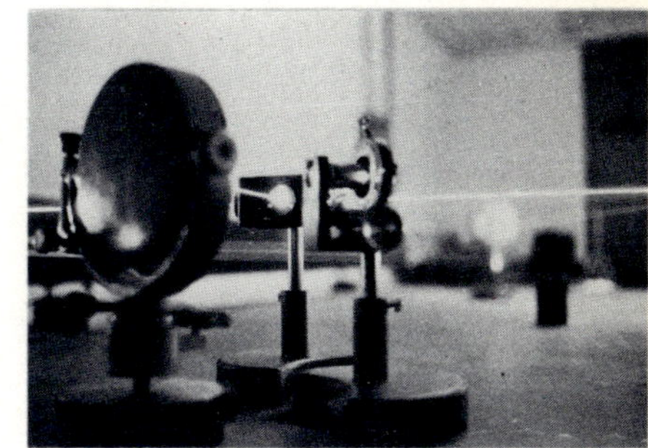

A beam splitter dividing beam into reference beam (left) and object beam.

Steering mirror redirecting beam.

But by illuminating this plate a three-dimensional image of the original subject can be produced, projected in space. Previously laser beams had to be used in the illumination of holograms but the discovery of improved photographic emulsions has made it possible for the holograms to be viewed now with less expensive light sources, such as mercury vapour, xenon or sodium vapour lamps or even by means of a tungsten halogen light bulb. Often if a laser beam is used the images appear almost too bright to view, and laser light also gives images a characteristic speckled effect.

The hologram can be viewed in two ways. If the plate is illuminated by the reference beam alone, from the same angle as it was recorded, a three-dimensional image is visible through the plate in an identical position to the original subject. This image is called the Virtual Image.

For the image to appear in front of the plate – a much more exciting phenomenon – a more detailed viewing process is necessary requiring the recording of a second hologram.

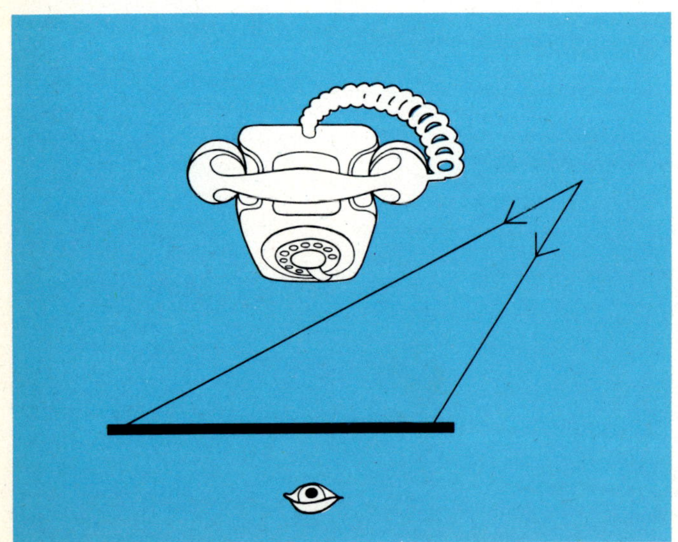

Viewing the Virtual Image. The reconstructed image appears in the same position as it was recorded, behind the plate.

Viewing the Real Image: in front of the plate but pseudoscopic (inside-out and back to front).

The original hologram is first illuminated from the back by a beam of diverging light. When viewed from the front, an image, known as the Real Image, is visible in front of the plate. This image, however, is back to front similar to the reflection in a mirror, and, because a hologram is a three-dimensional recording, also has a reverse perspective: that is to say objects in the background appear larger than those in the foreground. The correct scientific term for this image is pseudo-scopic.

To restore it to its original form the pseudo-scopic image is projected into space and used as the subject of a second holographic recording, for which a much stronger beam of light is required. This can be easily achieved by increasing the intensity of laser light.

The images viewable on this hologram will be the direct opposite to those on the first. The image formed on the far side of the plate when illuminated in its original position will be pseudo-scopic, while that projected in space between the viewer and the plate, an exact replica of the original subject. This true image in front of the plate is referred to as being orthoscopic.

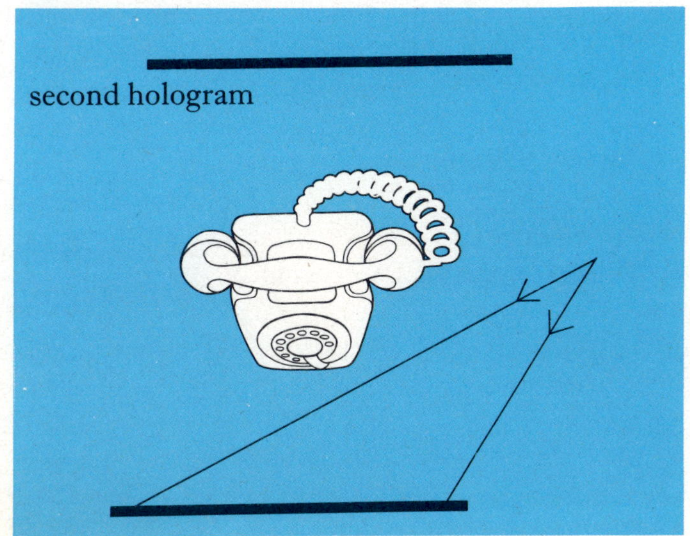

Recording the reconstructed image to make a second hologram.

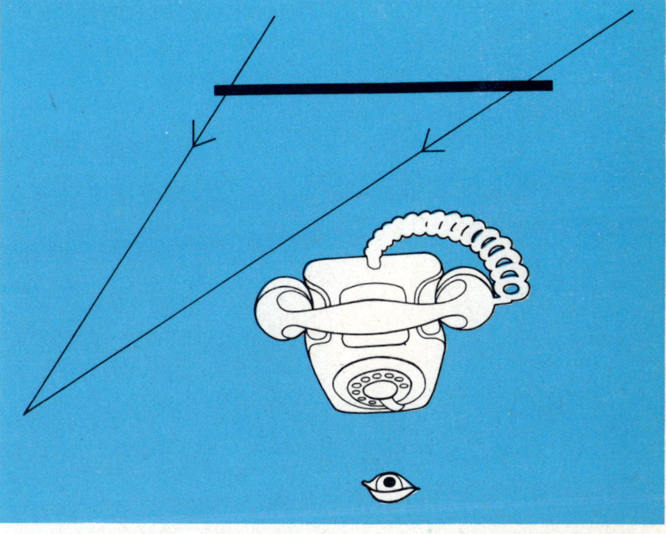

Viewing the Real Image of the second hologram which is now orthoscopic (the right way round).

It is, however, an expensive process on any large object or scene since the off-axis mirror required to collect the beam of diverging light is an extremely costly piece of optical equipment. To project in space a life-size hologram of a human being would require optics as yet only considered by astronomical observatories.

A large plate will achieve a more extensive viewing angle of a subject. If the plate is curved, for instance, it is possible to produce an image which, when viewed from the right angle, shows the top and bottom of an object. By using cylindrical plates or by constructing a square box out of four separately processed holographic recordings, it is possible to record a subject from all sides. By rotating the hologram or by moving round it, an image of the entire object can be viewed.

When taking a holographic recording the light waves from the subject are not focused to a point, as in photography, but allowed to spread out through the space between the object and the plate. This means that any part of the hologram contains a recording of that object from its own particular viewpoint. If the plate is broken, any one piece will retain the entire recorded image.

Another unique property of the hologram is that more than one image can be stored on any plate. By recording an image and then recording another image with the plate at a different angle to the reference beam, different images can be accepted. There is no limit to the number of images, apart from the practical problems related to a specific "set up".

It is often asked if full colour holography is possible. The answer is an unqualified yes, but colour holography is still in its infancy and much research work has yet to be carried out. In principle, it is possible by making a recording in the white light of a Krypton Laser, using a very thick coating of gelatine, to produce an image identical in size and colour to the original subject, but in practice the results to date have been far from perfect. The subject has to be placed close to the plate, rendering

An orthoscopic hologram of a telephone – made as described in the diagrams on the preceding pages.

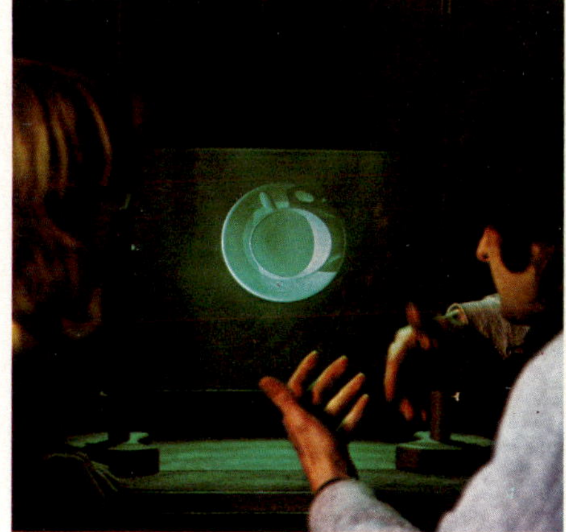

An orthoscopic hologram of a cup and saucer.

Both these holograms were recorded and constructed at Loughborough University.

the recording of sizeable objects impossible, and there is the problem that a number of images are being reproduced, so that the effective result is a blur.

Because of the differing wave lengths of all colours, holograms can vary in size on viewing. If an image is viewed by using the same colour light as employed for the recording, it will appear the same size as the subject. But should, for example, the recording be made in the blue-green light of an Argon-Laser, the image, if viewed in the red light of a Helium-Neon Laser, will appear larger than the original subject and nearer the plate.

However, if the same colour is used for recording and viewing, both the Virtual and the Real Image will be identical in size to the original subject.

One particular form of holography worthy of separate mention is the multiplex hologram, a method of storing photographic information holographically that is becoming increasingly common.

A series of still photographs or a motion picture recording of the subject under study is exposed, the number of stills or frames taken depending on the amount of view of the subject required on the finished hologram. After the film has been developed a series of slit holograms is recorded using each frame of film as a subject for each slit of holographic film.

The procedure can be totally mechanised so that a machine exposes a slit hologram for each frame of footage at a very rapid pace.

A multiplex hologram can either yield horizontal or vertical parallax depending on whether the camera is panned around the subject or craned over it. Flexible film coated with holographic emulsion rather than plates is most commonly employed.

The advantage of this type of hologram is that holograms can be made of almost anything that can be captured on ordinary film without the need of the expensive and clumsy pulse laser; but there is a disadvantage in that it is not truly a hologram, simply photographic information holographically stored.

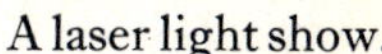

A laser light show.

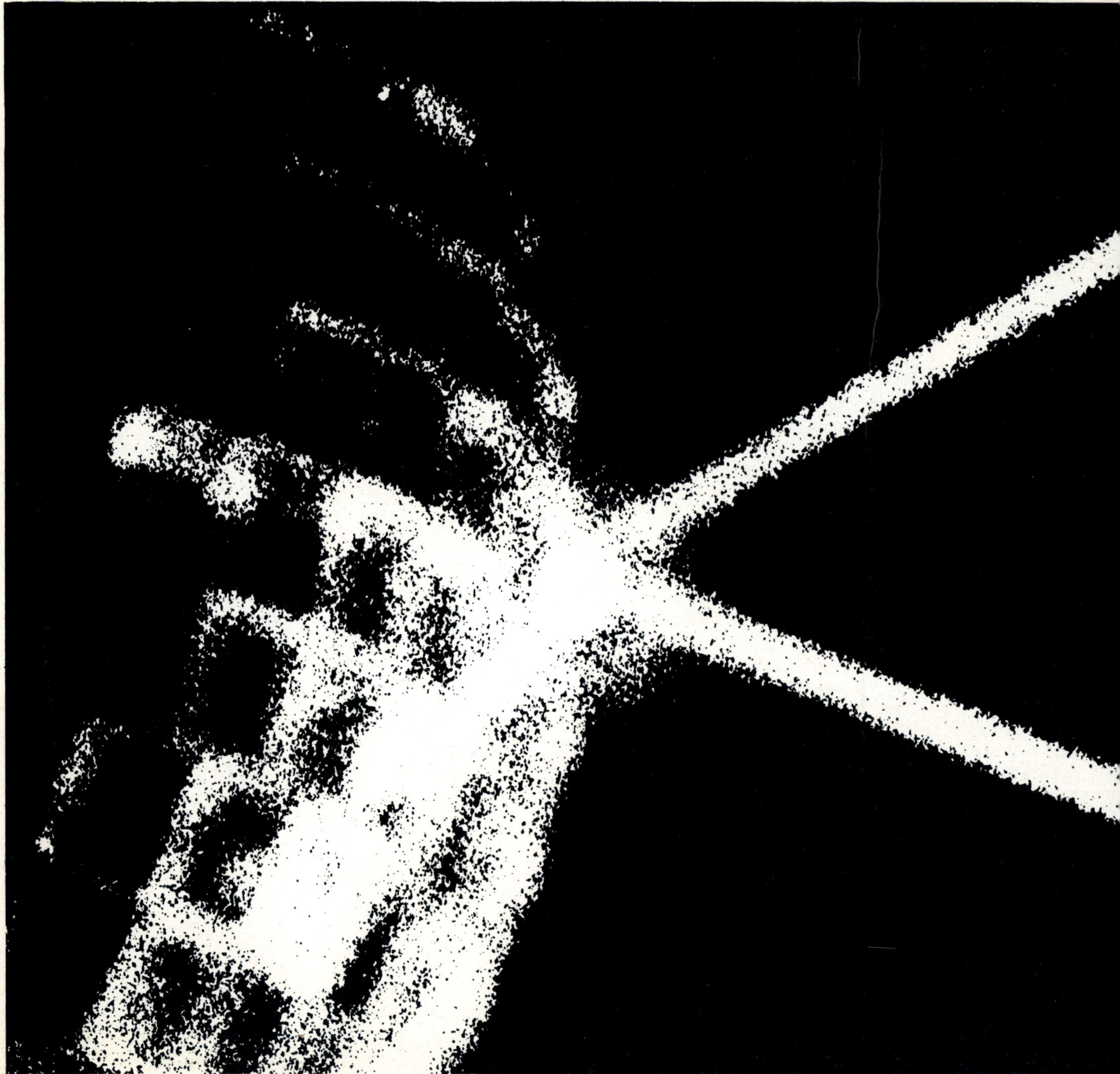

THE PRACTICAL USES OF LASERS AND HOLOGRAPHY

The "death ray" so ruthlessly exploited by the science fiction writers was at first thought to be the only use for the laser.

It is not possible to predict the eventual scope of lasers and holography, since the list of applications is constantly growing.

At first the laser was seen as a potent weapon, a "death ray" ruthlessly exploited by the science fiction writers. In the intervening years much research and development has been geared towards the laser's military potential, but while the SF writers' dreams of "vaporisation" may one day become a reality, many more constructive uses have been found for it.

Employed as a drilling tool a laser is capable at close quarters of blasting holes in hard steel and diamonds. It can be adapted to perform delicate eye surgery such as the painless removal of a tumour from the retina and the "spot welding" of detached retinas.

Wherever the need exists for small spots of intensely concentrated light, the laser is being tested and adapted, and breakthroughs often come from the most chance and unexpected directions. Scientists believe the day will come when the laser will provide beams of energy to allow satellites and space craft to receive power from sources on earth as they orbit about it or recede on their way into distant space. Ultimately it might be used as a means of communicating with other life forms – perhaps not human – on distant planets of suns other than our own.

Holography is already used for testing engineering structures, for visualising subtle aerodynamic flow and phenomena, and for the permanent storage of information at high densities. It has offered new insights into acoustical imagery, new possibilities in teaching and training aids, a new method of transporting and displaying priceless art treasures and a new sort of jewellery which picks up the surrounding light and glows. To physicists, holography, the most realistic form of stored imagery yet discovered, is one of the most exciting expressions of the laser.

A powerful argon laser beam hits the hand diffusing a flood of light on to the face. The heat of the beam becomes unbearable after a second or two.

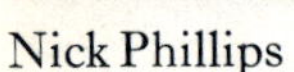

Nick Phillips

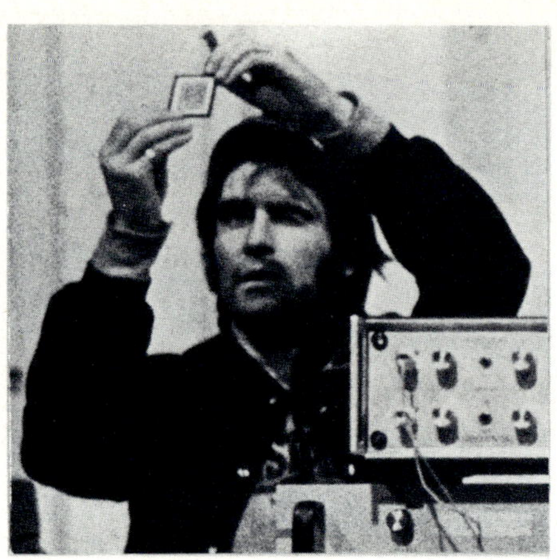

Anton Furst

John Wolff

A fine example of its multifarious uses lies in the fact that it originally brought together the authors, whose characters and interests most would consider relatively diverse. To Phillips, the scientist from Loughborough University, it is a fascinating science, very much in its infancy with horizons at the limits of his imagination; to Furst, the film technician, it offers the possibility of a revolution in the field of current filming techniques and in the whole area of visual effects in television, cinema and theatre; while to Wolff, the leading exponent of laser light shows, it is a source of endless fun and amusement which promises to add a new dimension to his existing work in the highly competitive world of pop music.

It was through the work of Wolff that The Who pop group first saw how laser light could be employed as a visual effect and in recent years they have been using lasers – the group now have eleven – at all their shows throughout the world. The very nature of rock shows demands the continual search for new and wilder attractions and as yet there has been nothing to match the mesmeric quality of laser light in helping to heighten audience involvement in the music. Future developments in holography can only add new dimensions to a medium already established as part of the present day rock scene. The cost of the equipment, however, restricts the use of laser light to the most successful groups.

If it is to retain its mystique it is imperative that those employing lasers do not become over-indulgent in their use of these new, expensive and sophisticated visual effects toys. For this reason The Who never use laser light for more than a few 90-second bursts during each performance.

The Who in performance. Photograph by Graham Hughes.

When a laser beam passes through smoke or cloud it appears to be solid.

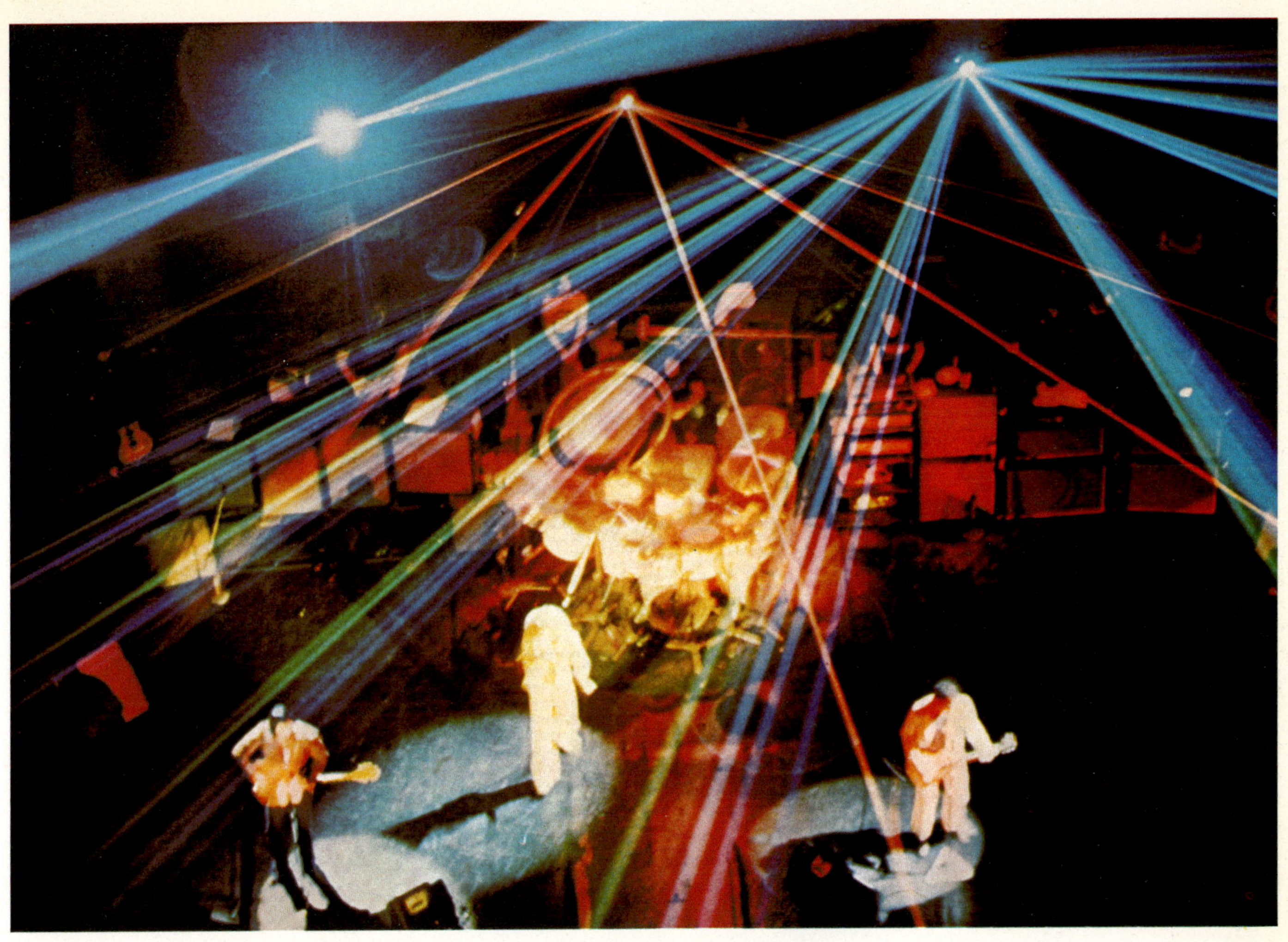

When split a laser beam can be made to create a multitude of shafts of bright pure coloured light that can in themselves be woven into patterns which, due to laser light's unique qualities, appear almost solid. The creation of dancing specks of coloured light caused by the reflection of beams from a silver ball placed strategically near the roof of the stage helps add to the hypnotic effect.

A ceiling of laser light.

Last summer Wolff put on a laser show in Venice with Paul McCartney and Wings as one of the events organised by UNESCO for the Venice in Peril fund. A mirrored ball in the roof of the Campanile diffracted dazzling shafts of coloured light over every part of St Mark's Square, thus highlighting the architecture of this beautiful city.

A laser beam split many times to achieve a fan effect.

The interwoven patterns of light can be raised and lowered in unison to create the illusion of a multi-coloured ceiling, and it is not difficult to persuade people into thinking that laser light is solid.

To the naked eye a laser beam appears to maintain a constant width. If a Continuous Wave Laser is directed at the moon at half meridian, the beam will have spread to 125 square miles on arrival; while in theory a collimated Continuous Wave Laser could project a parallel beam of light across the universe. A Pulsed Laser aimed at the moon yielded a two square mile patch of light, enabling scientists to read the distance from the moon to the laser to within a metre.

As the human eye is only capable of seeing a certain distance, a beam shone into space appears to come to an abrupt end. This is an illusion as anyone who tries to walk to the end of a laser beam will quickly discover, for however far one walks the beam never gets any shorter.

Laser beam reflecting off faceted mirror ball.

Light structures can be created in the sky by the use of oscillating mirrors placed in front of a laser. These need to be programmed by computer but always assuming the calculations are correct, there is no limit to the images that can be formed. Wolff has already achieved one of his ambitions by projecting words on clouds at open air rock concerts.

Lasers sign writing in the studio at Shepperton.

Smoke and clouds are essential ingredients of any light performance. Beamed through cloud laser light illuminates tiny particles which in turn reflect the light. The result is the creation of walls and corridors of light that appear to burrow deep into the clouds. Next perhaps it will be possible to illuminate a holographic image inside a triangular corridor of three-dimensional light.

Special effects with mist.

It is also not beyond the realms of fantasy that in the near future it will be possible to record a hologram of The Who in action. At a high point of their performance, after the audience have been aroused to fever pitch, all lights could be shut off and the hologram only illuminated. Then as the group played on hidden in the darkness of the stage, the eerie motionless holographic image could be slowly lifted up into the sky and the band made to seemingly float over the heads of the audience.

Moving fans of laser light.

Many tried and trusted but expensive and painstaking methods of filming could be made redundant by the development of holography. Travelling matt shots, glass shots and backcloths – all much used methods in film-making – could be made easier and the finished product vastly improved by the introduction of holograms.

At present two-dimensional backcloths commit directors to keeping their cameras absolutely stationary. Should backcloths be superseded by holograms, craning, panning and tracking could all be employed in these scenes for the first time, since the three-dimensional backgrounds would change perspective with the movement of the camera.

Painted glass plates are placed in front of the cameras when directors want to alter part of a set or location. Again the camera has to be kept stationary throughout the scene, any movement destroying the perspective of the substitute scene painted on the plate. Holography could be developed to bring about a much needed improvement in this film technique.

Although holographic cinema could be possible there is no suggestion that it will overshadow the power of the two-dimensional film as we know it. It is most likely to be valued as a useful tool in special effect film-making.

In the theatre it will be possible to project three-dimensional objects into scenes and to superimpose them over actual objects and people. A classic example of this would be Hamlet's ghost appearing at the flick of a switch. It will also be feasible to alter the perspective of sets simply by turning the hologram in front of the reference beam; and artwork for the theatre could, of course, be done to a small scale and projected up with true parallax and perspective.

Furst's design for a holographic auditorium gives an indication of the sort of entertainment centres that could be built in the future. The auditorium is designed to make full use of all the latest projection possibilities including holograms. The audience can be seated in the middle on a central podium, allowing for maximum freedom of performance as the podium can be raised, lowered and rotated by a hydraulic ram to ensure that the spectator's attention is properly orientated to the action taking place round him. Alternatively the central podium can be covered by a false floor and used as a performance area with the audience seated around the perimeter of the spherically-shaped building.

Michael York was the first actor to perform in a hologram when a multiplex recording of his head appeared in the film "Logan's Run". The film is about life in the 23rd century but by the time that is reached, holography will have long since become a familiar phenomenon.

Furst's Picture Palace. Designed to incorporate maximum flexibility and full use of the most recent visual techniques.

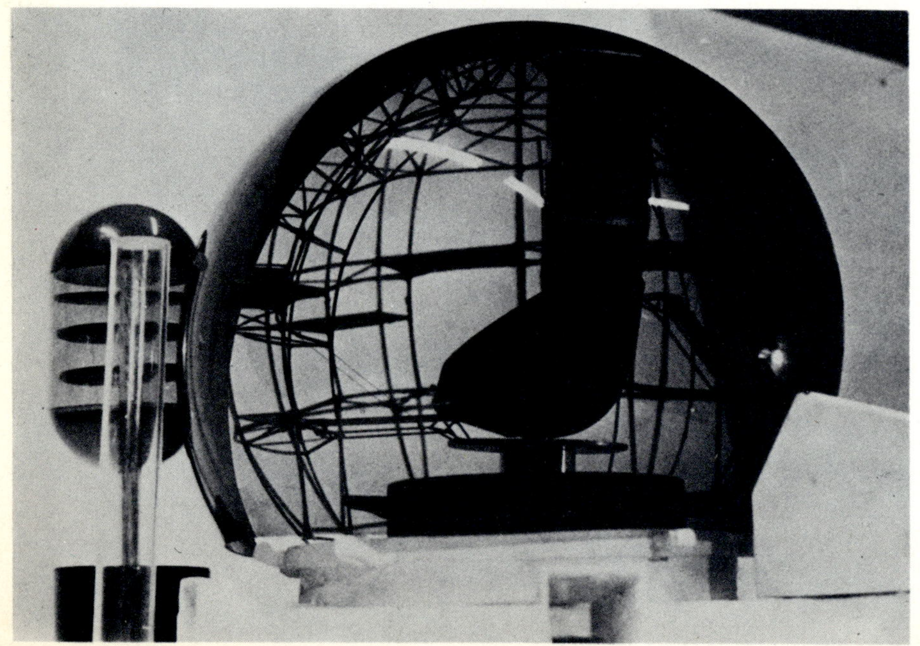

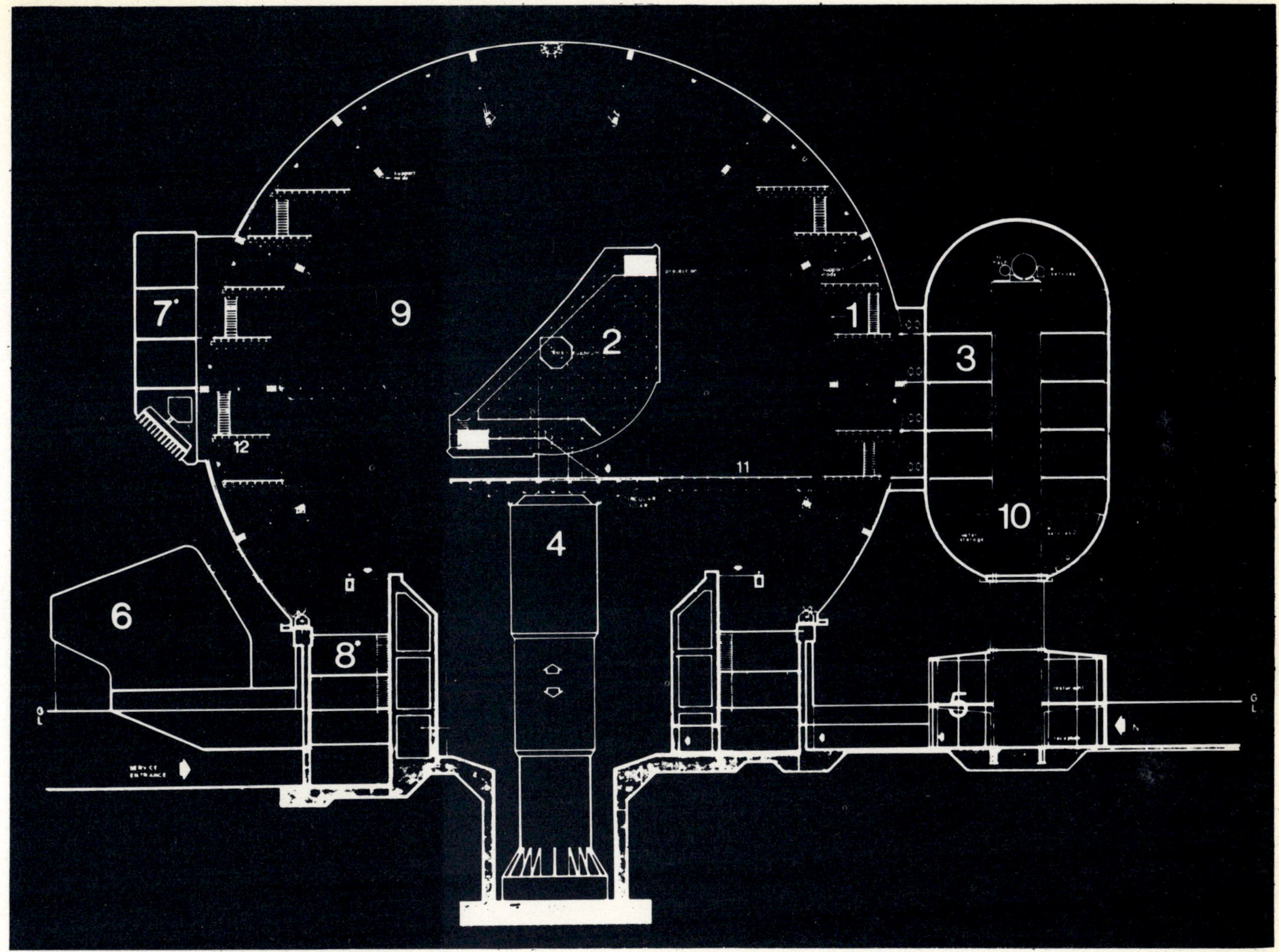

Section through structure showing
1. Main structure with support nodes
2. Podium carrying seating for audience
3. Ancillary public areas
4. Hydraulic ram
5. Reception and public facilities
6. Motor room
7. Control room
8. Annular wings
9. Performance space
10. Lift shafts
11. Access drawbridges
12. Cavity for back projection, equipment, acoustic shields, etcetera, or as seating area as described in text

Structural engineer: Anthony Hunt Associates   Hydraulics by Vickers Limited

Our show at the Royal Academy is therefore merely a demonstration of the early stages of a new skill.

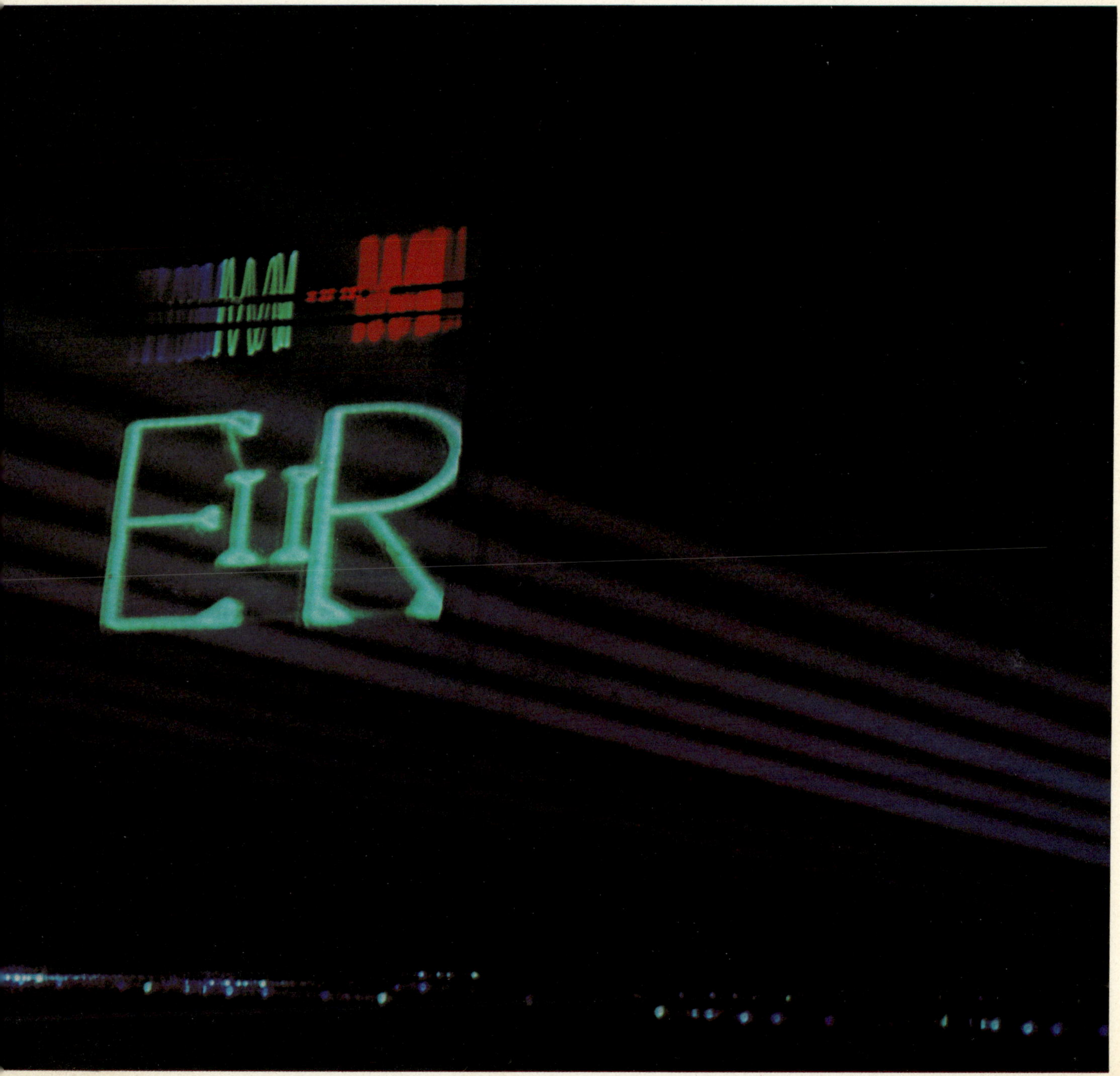

This book was co-produced
with Holoco Limited
for the exhibition
light fantastic
at the Royal Academy in 1977

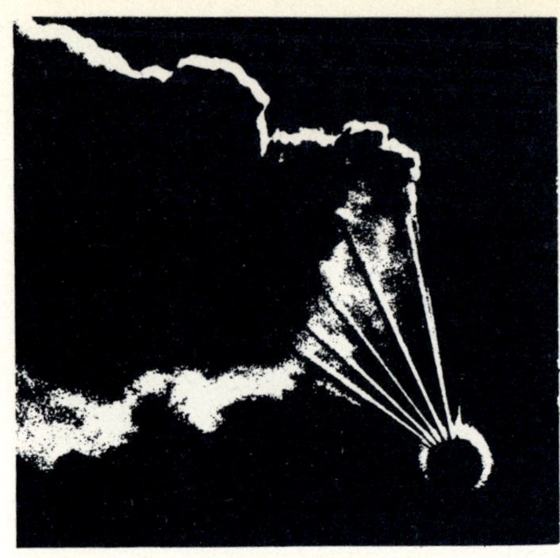

Designed by Tamasin Cole
and Theo Bergstrom

Diagrams by Kate Canning

Photographs by Theo Bergström

Exhibition designed by
Douglas Stephen.

Production services by
Book Production Consultants,
125 Hills Road, Cambridge

With thanks to
The Who
Neil Irwin & David Watling

David Porter, Ralph Cullen
The Physics Department
of Loughborough University

Gerald Leitch, Bernard Hunt
Chris & Dave Matthews, Cyrano
ML Executives, Clare St John

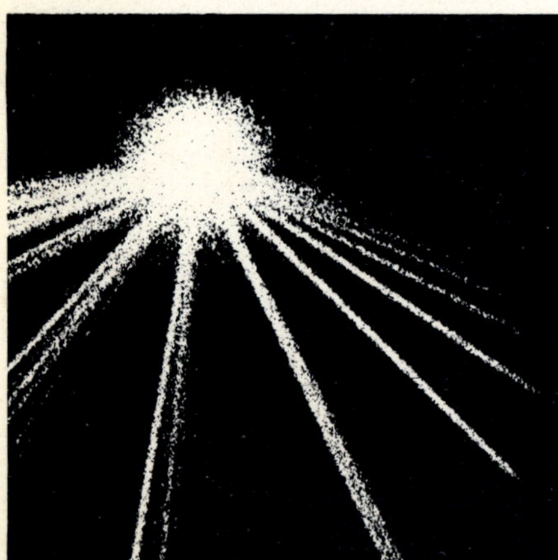

Produced and published by
Bergström + Boyle Books
Limited
22 Maddox Street,
London W1R 9PG

Made and printed
in Great Britain by
The Sackville Press,
Billericay, Essex

 ISBN 0 903767 17 1